Genesis – Rock Solid

A Biblical View of Geology

By Patrick Nurre

Genesis – Rock Solid
A Biblical View of Geology

By Patrick Nurre

Genesis – Rock Solid
A Biblical View of Geology
Copyright 2009 by Patrick J. Nurre
Second edition 2014
Published by Northwest Treasures
Bothell, Washington
425-488-6848
NorthwestRockAndFossil.com

All rights reserved.
Printed in the United States of America. No part of this book may be reproduced in any manner whatsoever without written permission except in the case of brief quotations embodied in critical articles and reviews. You may request permission in writing by way of email at northwestexpedition@msn.com.
Scripture quotations taken from the New American Standard Bible®, Copyright © 1960, 1962, 1963, 1968, 1971, 1972, 1973, 1975, 1977, 1995 by The Lockman Foundation. Used by permission. (www.Lockman.org)
Cover photo: The view of Yosemite Valley from Tunnel View in Yosemite National Park, California, United States. By Diliff - Own work, CC BY-SA 3.0, https://commons.wikimedia.org/w/index.php?curid=26413820.
Title page: Grand Canyon of the Yellowstone, Yellowstone National Park. Photo by Patrick J. Nurre.

Genesis – Rock Solid
A Biblical View of Geology

Contents

Preface	4
Introduction – My story	6
Chapter 1 The State of the Church Today	12
Chapter 2 From There to Here!	14
Chapter 3 Uniformitarianism vs. The Biblical Account	25
Chapter 4 Blinded for 200 years – The Story of J Harlan Bretz	38
Chapter 5 The Geology of Genesis – Putting on the Right Glasses	44
For Group Discussion	55
Bibliography	56
Picture Credits	57

Preface

Geology is an extremely broad subject! Some find it absolutely boring while others, like me, are fascinated with the rocks and fossils all around me. There have been thousands of books written on these things and personal opinions and ideas have been just as numerous. Likewise, many people have written great books discussing the evidence for a young earth and global flood. However, I have seen very few books clarifying the historical development of the philosophy that under girds modern geological thought. I personally believe this is a weakness in the modern church. Because of this I think believers in a straightforward reading of Genesis are confused by the apparent scientific evidence.

This booklet is an attempt to help the average person to get a simple overview of the fascinating history of the development of modern geology and begin to view the earth from a different perspective. In my experiences with teaching Genesis and geology over the years I have found that there is a need to equip those who embrace a straightforward reading of the Book of Genesis with the following: (1) Discernment in recognizing the difference between the philosophy of science and the facts of science. On the surface this may appear to be obvious. But it is not. Modern science is a mixture of both testable and untestable ideas. It is often difficult to see these differences. (2) Viewing the geological evidence with a Genesis perspective or interpretation; a set of glasses, if you will – a straightforward reading of the text. With these things and a little bit of general knowledge about rocks, one can begin to see the geology of the earth in a new light.

This short booklet concentrates on exposing the philosophy underneath modern geology, how it took hold of modern scientific thinking, and where it conflicts with the Scriptures. You can follow up your introductory reading with four other books which get into the geology much more in-depth. Those books are:

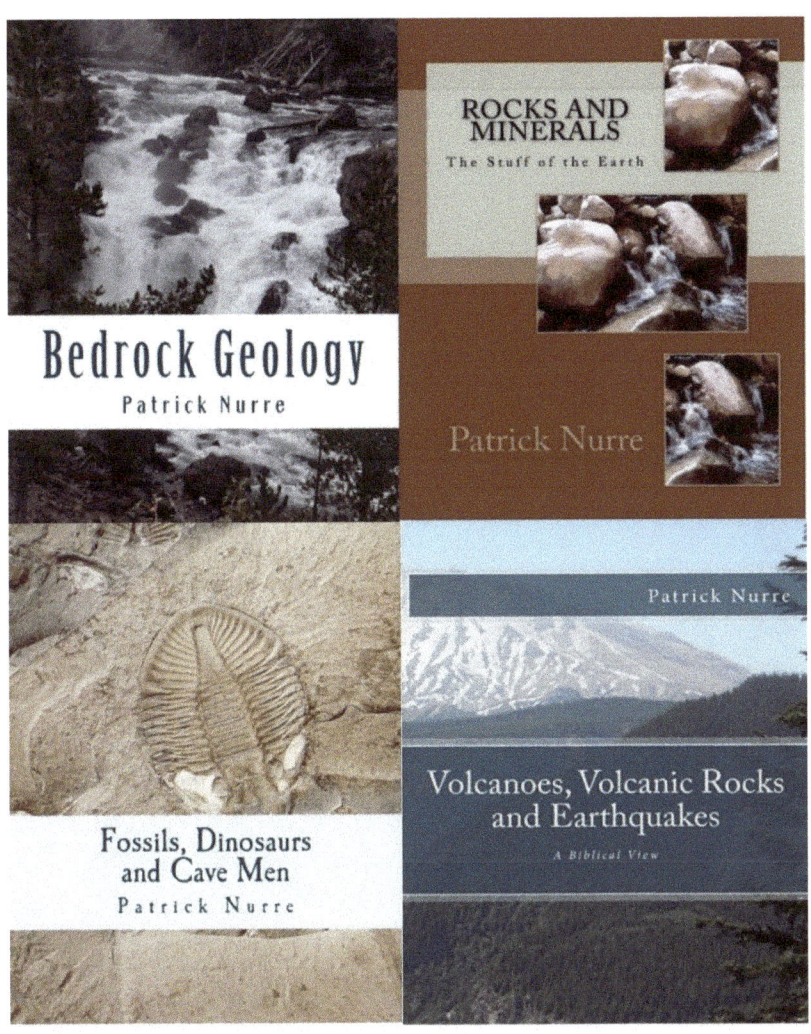

These books are available from Northwest Treasures
www.northwestrockandfossil.com
425-488-6848

Introduction
My Story

I grew up in a very cool state – Montana. How many people do you know who are from Montana? We are a treasured few! I lived just a stone's throw (no pun intended) from Custer Battlefield and the old Fort Custer. I spent my weekends digging for 7th Cavalry treasures and collecting fossils from the bluffs of the Big Horn and Little Big Horn Rivers. What an exciting time for a young boy!

The rugged Montana I grew up in, along the Big Horn River

As with most young children today, I was fascinated with rocks as a kid. The area I grew up in is rich with geological features and fossils – The Big Horn Mountains and Big Horn Basin. I took every chance to collect and admire the precious treasures of these beautiful places. I was probably the only junior high kid to hang out with the Montana Historical Society – a large group of old women (to me they must have been at least 100 years old) who knew the secrets of the high prairie. By the time I was in high school, I had an extensive collection of interesting artifacts, rocks and fossils that I had collected on many of these trips.

My mom and dad enthusiastically encouraged their young scientist son. Truth be known, they were probably glad that I was not getting into trouble like several my peers. My dad would drive me and another friend out to a deserted area of the Big Horn bluffs and drop us off. All day long we would dig for marine fossils, watch for rattlesnakes and listen to the Big Horn River as it gently flowed by our location. What fun days!

The Big Horn Canyon, just 40 miles from where my home was: I hiked this canyon many times as a young child.

As with most families in those days, my parents were careful to instill certain moral values in me. I grew up in a large Catholic family, devoted to the Catholic Church and its teachings. As a young Catholic boy, the nuns impressed upon me the Biblical thought that Adam and Eve were our first parents. I was taught that God had originally created man and woman and that they fell through disobedience to God. I was also taught that Noah was a real person and that the Flood of Genesis destroyed the earth and the sinners.

During these young formative years, I spent hours at the local Carnegie Library, listening to the banging steam pipes, enjoying the smell of old books and devouring every book I could about geology, fossils and dinosaurs. Little did I realize, however, that along with my scientific curiosity I was being indoctrinated into a particular viewpoint that was not science. Almost without realizing it, my religious beliefs were being torn down and replaced with something else. Slowly but surely my belief in Adam and Eve, the Fall and the Genesis Flood were replaced with an evolutionary view of life. And although I struggled with this for a while, by the time I entered college, I found myself mocking Noah and his tiny boat as just a myth. My new intellectual belief in evolution, long ages of earth history and chance processes was exciting. I believed that I understood things that my peers did not.

While attending college, something else was happening inside of me. I was beginning to face questions that I had not thought much about before. My new-found faith was not answering the nagging questions of, "Who am I?" and "Why am I here?" Try as I might, I could not escape from my feelings of purposelessness. My evolutionary view of life was not furnishing any satisfying answers. To make a long story short, a friend directed me to the Bible for answers. When I finally realized that my problem was one of rebellion against God, I asked God for forgiveness and started my new life in Christ.

Shortly after becoming a Christian, I found myself in conflict with what I thought was science. I was ashamed of the mocking I had done about Noah and his boat, yet I could not ignore the geology I had studied through the years. I had to resolve this issue in order to move forward in my Christian life.

About this time, I was given a book that was to open my eyes to the real issue. That book was *The Genesis Flood* by Henry Morris and John Whitcomb. Morris and Whitcomb looked at geology from a different yet reasonable perspective – the geological implications of the global flood as told in Genesis. As I read it I found myself getting angry at the geology establishment that I had believed in for so many years. Why would geologists withhold such valuable information from me? I began to realize that much that passed for true scientific discovery was better labeled as belief, not science. Dates for rocks and fossils were being interpreted based on a belief in a system, not on scientific foundations. From that point on I took a new look at the rocks I was collecting and studying. Before coming to conclusions, I would ask myself the question, "What does the Bible say about this?" My interest in geology and in earth history has been a wonderful journey since then.

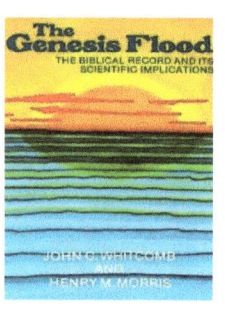

In today's world, the modern view of geology has replaced the Bible as the source for understanding earth history and humanity. Many churches share this position. We desperately need to reevaluate what has been our cultural position on earth history for the last 200 years. Our modern church age is locked internally in an agonizing struggle just as I was as a young believer. These issues can divide and leave a state of unrest and anger among Christians. But we cannot ignore these explosive issues; they will not go away. And the price for abandoning the Scriptures in favor of contemporary thought is too high. We must pray and seek instruction from God's word as the church has done for centuries.

Throughout its history, the church has exhibited an amazing quality – the ability to sort through problems, and to reform itself. The church has gone back to the Scriptures and started with the basic belief that their Holy Bible was a revelation from God and that it would provide the answers to life's perplexing questions. Although penned by men, the conviction has always been that the Scriptures were the very words of God communicated by God to mankind.

I write this booklet with the conviction that the Scriptures are the very words of the eternal God and as such are timeless – not bound by cultural influences or thinking, however advanced. It is also my conviction that God has protected the integrity of His word through the ages as it has been translated and passed down to us from His prophets and His apostles. So, this booklet is not a defense of the Scriptures, but a defense for using the Scriptures to interpret the rock, mineral and fossil record around us.

Geology is a combination of two Greek words that mean the study of the earth. When we study the earth, as with anything we study, we must choose a starting point. The rocks and landforms of the earth do not come with chronological commentary or dates stamped on them. A starting point that comprises one's worldview interprets the rocks.

CREATION	EVOLUTION
light created before sun	no light before sun
oceans before land	land before oceans
first life was land plants	first life began in water
trees before fish	fish before trees
plants before sun	plants after sun
stars after earth	stars before earth
birds before reptiles	reptiles before birds
man from dust	man from previous animal
man before rain	rain before man
man before woman	woman before man

Both presentations tell a story. They cannot be harmonized. Neither one can be proven by science, for science cannot test the unrepeatable past. Both are worldviews.

In the sometimes-perplexing issues of geology, I have chosen the Scriptures as my starting point. As I stated earlier, this was not always the case in my life. By early adulthood I had embraced the belief that earth's history was one of millions of years of slow change and a series of processes purely of chance. But this view did not come about as a result of examining the rocks, but by embracing an interpretation of the rocks – one of millions of years of slow and gradual change. I simply believed what others who were supposed to know, told me. That became my starting point. As a result, I rejected the Bible stories of Adam and Eve and Noah's Flood. What is your starting point?

As Christians, we believe the gospel of salvation because God revealed it. Isn't that what faith is all about? Doesn't faith trust that what God has revealed about anything is the truth? As Christians, we are commanded by God to live by faith. So, the starting point in any controversy should be, "What does the Bible say?" Even if evolutionary dogma should tempt us to think otherwise, we must first find out what God might say about an issue. If the Bible is true in all that it states, then we should expect the world around us to match up. The evidence of the rock record should make sense in light of Scripture. The modern trend has been to reinterpret the Bible in light of what modern man has told us to believe about the earth and its history. This is dangerous and should be rejected quickly in our search for understanding.

I trust that this booklet will help you in answering some of the tough challenges we face in the field of modern geology. If you wish to correspond with me, I would be thrilled. You can reach me at northwestexpediton@msn.com. It is such a joy to get out into nature see God's creation and understand it from a Scriptural foundation.

One other thing, if you are ever in the Seattle area, look me up. I think you just might enjoy a geology field trip in Washington State from a Biblical perspective.

Mt. St. Helens after the eruption, part of the Cascade Range of Mountains

The Cascade Range of Mountains, a combination of volcanic and metamorphic rocks

Chapter One
The State of the Church Today

Did you realize that geology, the study of the earth, is one of the major battlegrounds in the church in the 21st century? It is not the science in geology that produces the conflict, but the worldview espoused by modern geology. That's right. Modern geology teaches a different history of the earth than does Genesis. It is in actuality an opposing view to Genesis, but packaged in scientific language. This confuses many people who want to believe the Bible, but are overwhelmed by the apparent science which seems to contradict the Bible. There is a great deal of conflict generated at this point. Many have tried to resolve this by concluding that how one views Genesis must not be that important or doesn't really matter that much. But it does matter, and for one very important reason: The Lord Jesus treated Genesis as true in its entirety.

Jesus quotes from the Book of Genesis more than any other in the Old Testament. He saw Genesis as true history as He used actual people and events such as Adam and Eve and Noah and the Flood to teach important truths. In Matthew 4:4 He stated, *"It is written, 'Man shall not live on bread alone but on every word that proceeds out of the mouth of God.'"* This included the creation in Genesis chapter one and the Flood in Genesis chapters 6-9! If Jesus cannot be trusted to handle that which was revealed in the Old Testament, then why should I even pause to trust Him for my salvation – revealed in the New Testament?

In our schools and in the media (even in some churches!) we are taught as certain fact that the history of the earth and the universe has been a long one spread over billions of years. We have been taught that the fossils definitively tell us of an evolutionary parade of life as it struggled for existence. We are taught from youth up that the Bible, while a great moral book, is either patently false when it comes to earth history or is full of allegories.

If these allegations are true, then we must face a host of questions. Can we trust the Bible when it comes to questions about life and morality if it is wrong in its presentation of earth history? Should we face the fact and acknowledge that the scientists have this one right? In the midst of this relentless onslaught, many in the church have wavered in their faith and commitment to the Scriptures and have compromised their view of a six-day creation and a worldwide flood. We try to make sense of Genesis in light of the so-called findings of geologists but are not equipped to do so. We conclude that the Bible must mean something else in the first eleven chapters of Genesis.

The church by and large has grown comfortable with relegating the issues of a young earth and global flood to irrelevant status, believing that what the Bible has to say about this area either is unclear or does not matter. And we go on defending the gospel as if this issue does not matter. But if it does not matter, which part of the Bible does matter, and why should we believe that it does? What part of the Bible can we defend as being trustworthy? Can we really defend it when it comes to our eternal destiny?

The predominant Christian view of today advocates that we as Christians rely on the discoveries of enlightened scientists for the real meaning and interpretation of Genesis chapters 1-11. This view says there are two revelations of truth: The Bible and the world of nature. Christians are to come up with the spiritual meaning of the Scriptures. But the domain of interpreting the physical world belongs to science. When it comes to understanding earth history, Christians are to yield to science for a correct understanding of earth history. This idea is totally inconsistent and absurd for the Christian who says he or she believes the Bible. If the Bible is distorted and unintelligible as to its record of Jesus' resurrection, how are we to know the difference? How can we live such a dichotomous faith (a faith consisting of two contradictory ideas)? We believe that the Bible is the word of God, yet for part of it, at least, we don't trust what it seems plainly to state – that the earth and all it contains was created in six 24-hour days (Moses had understood this from Exodus 20:1-10) and that there was a global flood that destroyed the original landforms. This was the view of the Lord Jesus. We may agree that the straightforward reading of Genesis conveys a young earth and global flood, but we yield to modern geology as our interpretive guide, because after all, we live in a modern, enlightened, scientific world.

How did we get into this predicament in the first place? How did the church, which for centuries had trusted the plain teaching of Scriptures on earth history, abandon this position and adopt a different view? Was it really because of the vast amount of scientific evidence that had been accumulated? Were there other factors that contributed to the demise of a young earth and global flood view?

Chapter Two
From There to Here!

Many today do not realize that evolutionary thinking started in the late 1700s with a particular philosophy of earth history, not with biology. Charles Darwin's *On the Origin of Species* (1859) was written after the idea of an old earth was firmly in place in the field of geology – by 1830. The developments in modern geology drove the development of modern biology that expanded greatly after Darwin.

The main battles within the church today are not fought over the issue of evolutionary biology. Most Christians today think that the idea of frog to prince is a silly one anyway. No, the real battles within the church are being fought over the age of the earth and the legitimacy of a global flood – the domain of geology! My observation has been that the majority of the modern church now drifts toward the side of secular geology when they want answers to questions such as, "How old is the earth?" and "Was there a worldwide flood?"

Oddly, it was clergymen who first began to rely upon science and geology to furnish us with the answers to the age of the earth and the formation of its landforms. Notice the following examples:

• One of the earliest of clergymen to preach that the earth was millions of years old was a young preacher named Thomas Chalmers (Free Church of Scotland, 1780-1847). He is credited with creating the **Gap Theory** – that the long ages of earth history occurred between Genesis 1:1 and 1:2. This was intended to be an encouragement to the church, but in reality, was a compromise position, harmonizing the Biblical position with secular geologists' views. Although highly imaginative, it is rather a ridiculous idea, as there is absolutely no evidence, Scriptural or otherwise, for this view. Yet it has influenced millions of Christians to compromise and to embrace the position of modern geology.

• One of the key figures of the early 1800s was an extremely influential Episcopalian clergyman and geologist by the name of William Buckland

(1784-1856). He was Dean of Westminster and a member of the Royal Society. He tutored many of the scientists of the 1800s, including Charles Lyell. Lyell was perhaps the most influential person in modern geology. Buckland became outspoken in his belief that the book of Genesis was not to be taken literally and that the earth was formed over very long ages.

William Buckland

• Other clergymen included Anglican theologian, George Faber (1773-1854) who formulated the **Day/Age Theory** (the day in Genesis is equal to millions of years), and Congregationalist and geologist John Pye Smith (1774-1851) who argued for a localized and tranquil (no worldwide upheaval or catastrophic geology) flood of virtually no geological significance. This of course is quite a bit different than the picture of a violent upheaval of earth's foundations as painted in Genesis, as we will see in chapter three. These two views are still advocated today by many evangelicals as reasonable explanations of the Scriptures in light of the findings of modern geology.

• Charles Spurgeon (1834-1892), that great Baptist preacher, accepted an old-earth view of geology.

- I have personally used the commentaries by the great Bible scholars, Charles Hodge (1797-1878) and his son A.A. Hodge (1823-1886). Although these men adamantly rejected evolutionary theory, they advocated an old earth view based, not on the Scripture, but on the findings of secular geologists.

- B.B. Warfield (1851-1921) in the late 1800s followed A.A. Hodge as lead theologian at Princeton University. Curiously, Warfield and Hodge aggressively defended the inspiration of the Scriptures in the late 1800s. But, throughout his career, Warfield went back and forth on his views of Genesis. Many historians have labeled him as a *theistic evolutionist* which means that God created everything and then let it evolve.

- C.I. Scofield (1843-1921) probably did more to encourage the reading of the Bible in the first part of the 20th century than any other man through his *Scofield Study Bible*. Yet, this well-known Bible scholar advocated the Gap Theory of Thomas Chalmers (mentioned above) when it came to interpreting Genesis chapter one.

- Closer to home, the late Gleason Archer (1916-2004), apologetics giant to the church, believed that although a straight-forward reading of Genesis gave the picture of a young earth and universe created over six 24-hour days, he expressed that this kind of reading of Genesis ran counter to modern scientific research. He believed that the earth and universe were billions of years old.

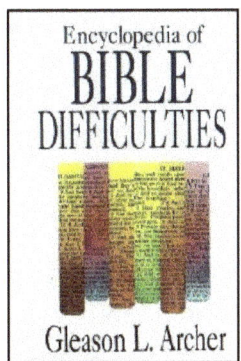

Gleason Archer

- InterVarsity Press, for years known as a publisher of fine Christian literature, publishes the works of Hugh Ross (b. 1945), an astronomer, who, although a Christian, advocates an earth and universe billions of years old. His view of the flood in Genesis is that it was a local, tranquil flood of very little geological significance. Hugh Ross is highly influential with many evangelicals today.

The list goes on and on. These men compromised, not because of a lack of clear teaching from the Bible on this issue, but rather despite it. They fell victim to pressures from what were supposedly scholarly, scientific facts. These "scientific facts" were really philosophical shifts developed by esteemed intellectuals. Such preaching and teaching from highly influential Bible scholars has gained a following against those who believe that Genesis teaches a young earth. Is it time for us to admit our error and join their numbers?

How We Got Here from There – A Historical Perspective
One of my favorite subjects is history. I love reading about people and events of the past. There is so much insight that can be gained from what other people have experienced. The historical development of geological thought is one of those fascinating and enlightening pieces of history that has been largely neglected and consequently forgotten in modern study.

Why, after a few thousand years, did the church change from its position of belief in a six-day creation and a global flood to one of a gradual unfolding of history of millions of years and no global flood? What factors influenced this change?

Most historians readily admit that the church has long embraced a six 24-hour day creation and a global flood. The teachings of Martin Luther, John Calvin and John Wesley are very clear as to this position in their commentaries on Genesis. All three clearly taught that the earth was young and that there was a global flood of significant consequence. During the 1800s, in the midst of scientific upheaval, there was a significant group of men called the Scriptural Geologists. These men adamantly opposed the drift of geology into an old earth paradigm. They wrote many scholarly and brilliant papers in support of geological reasons for a young earth and a global flood. They foreshadowed the modern creation movement in geology. For an excellent treatment on the history of these men, read the book, *The Great Turning Point, The Church's Catastrophic Mistake on Geology,* by Terry Mortenson. Most have never heard of the Scriptural Geologists let alone what they stood for. Rather, the history we have been told is that science freed itself from church doctrine and brought to light the truth of scientific evidence. Was it really this way?

 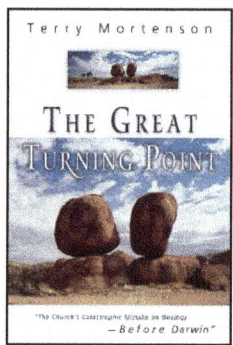

Terry Mortenson

In the late 1700s and early 1800s, there was a host of men who brought about significant changes in geologic theory. For the sake of space, I want to focus on just two who shook the way western culture looked at geology – James Hutton (1726-1797) and Charles Lyell (1797-1875). The cultural setting for these momentous changes that culminated by the mid-1800s was The Enlightenment (1600s-1800s).

James Hutton and Charles Lyell

The Enlightenment

The Enlightenment brought about a philosophical shift in the way man began to look at the world around him. The Enlightenment has been called The Age of Reason. In *The Church and the Age of Reason 1648-1789*, author Gerald R. Cragg wrote,

> *The latter 17th century was a period of rapid scientific advancement. Interest in the study of nature was general, and most European nations contributed to the expansion of knowledge…but the impact of new knowledge on old patterns of thought was raising questions that religious thinkers would have to face. A new class was beginning to emerge: men of skeptical outlook, impatient of all restraint…God Himself was expected to produce credentials satisfactory to human reason…it was not a doctrine about religion but an approach to its problems.*[1]

The Enlightenment has also been called The Age of Revolution, based on political revolutions occurring during the period. The revolutions were not just against oppressive governments, but also against God. This was made abundantly clear during the French Revolution. Instead of looking to God, the Scriptures and the church for answers to man's existence and meaning, the revolutionaries turned to human reason for their answers. During the Enlightenment, the church was no longer looked to by the intelligentsia for what to believe about the history of the earth. Man and human reason were crowned the new god. The Enlightenment was a period of time when the church was no longer looked to by the intelligentsia for what to believe about the history of the earth. Man,

---

[1] Cragg, Gerald R., The Church and the Age of Reason 1648-1789, 1970, Penguin Books.

through his highly-developed reason and diligent study of the earth could discover the secrets of the past on his own. God, the Bible and the church were unnecessary. This religious outlook became known as Deism. Deism is a belief in God based on reason rather than revelation, and it involves the view that God has set the universe in motion but does not interfere with how it runs.

James Hutton
One of the most influential men in the development of geological thinking during the Enlightenment was James Hutton (1726-1797). Hutton has been called the "Father of Modern Geology." He was a Scotsman who trained professionally as a doctor, but loved the field of geology. Hutton was a Deist, and most Deists believed that although God may have created the earth, He was no longer involved in maintaining it.

James Hutton

Deists generally denied the existence of miracles. Therefore, observable landforms were a product of natural laws and endless time. Some Deists believed in a Creator, some did not. Some were Unitarians in their church affiliation, others were Anglican or Presbyterian. The important point is that Deists taught that the earth should be studied apart from any influence from Scripture or the church.

Few realize that it was Hutton's Deistic belief that influenced his geologic theory. His conclusions did not come from his great collection of rock and fossil evidence. In his momentous work entitled Theory of the Earth, published in the late 1700s, Hutton concluded poetically that the Earth offered, *"...no vestige of a beginning – no prospect of an end."* The history of the earth was cyclical. As it was still early in the development of modern geology, many scientists of his day branded Hutton an atheist for his conclusions. This accusation mentally plagued Hutton till the day of his death in 1797. He nevertheless left his mark on geology. Because he saw neither beginning nor ending to earth history, he advocated a view of earth history called ***uniformitarianism***. This term espoused the thought that the earth's geological history could be explained in terms of slow, gradual and repeatable processes, processes operating in the present, without resorting to a Creator or global flood for an explanation. So, a river over a long period of time would carve a canyon. Mountains would rise over time, and so on, all without resorting to a global flood. If the history of the earth could be explained in terms of slow and gradual processes operating over long periods of time, then there just simply was no more need to invoke the global flood of Genesis to explain earth's geological past. Hutton, the Father of Modern Geology, ushered in a brand-new direction, a sharp break from Scriptural mores. Modern geology got its idea of the Rock Cycle from Hutton's influence.

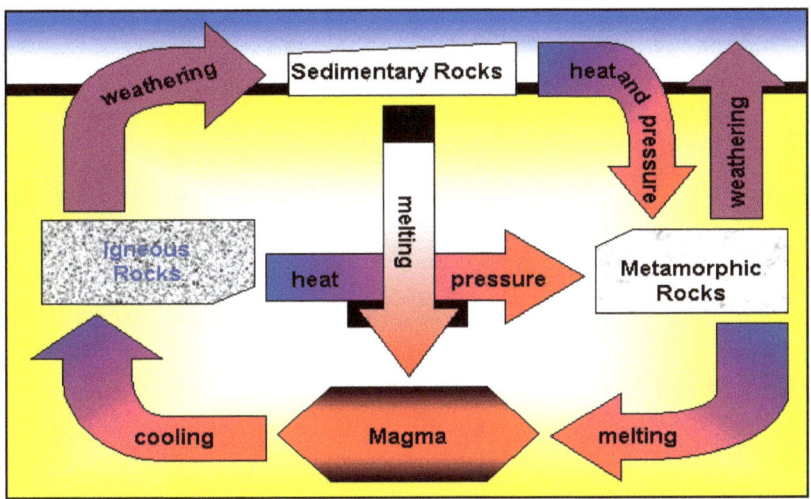

The Rock Cycle: Where is the beginning and the end? The rock cycle illustration gives the impression that the earth has developed by on-going repeatable geologic processes, which would exclude a global catastrophic flood!

Charles Lyell

The second most influential man in the development of modern geology was a man named Charles Lyell (1797-1875). Lyell was a Scottish Lawyer turned geologist. He was energetic and eloquent. Like Hutton, he was a Deist. Charles Lyell is credited with popularizing the phrase,

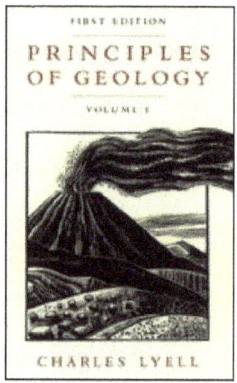

uniformitarianism – the present is the key to the past. His major contribution to geology was his three-volume work entitled, *Principles of Geology* (1830-1833). It is still in print and studied today!

It was volume one of this book that young Charles Darwin brought with him on his famous voyage as naturalist on the HMS Beagle from 1831-1836. Darwin himself credits this book with his own enlightenment in understanding how uniformitarianism applied to biology. The pieces of the puzzle for biological evolution came together for Darwin largely because of reading Lyell's book.

Young Charles Darwin and the Beagle – the 1830s

Charles Lyell, more than any other geologist, influenced the shift away from a belief in Genesis during the 1800s. Even his geology professor, William Buckland, once a believer in the Genesis account, publicly converted to a uniformitarian view of earth history because of Lyell's arguments. However, few realize that Lyell had a personal and passionate prejudice against Genesis—he disdained any reference to Moses. Publicly he was careful about criticizing the Genesis record of a global flood because the general public still believed in the Scriptural account. Privately his letters were caustic and scathing. His personal goal was to

eradicate once and for all the influence of Genesis on geology.

The doctrine of strict uniformitarianism held sway in geology until the late 1970s – over 140 years! In the 1970s a revival of sorts sprang up among many geologists in recognition of the evidence that catastrophic events played a significant part in the formation of earth's geological past. This movement was called Neocatastrophism. Initially the scourge of modern geology, it is now accepted as a legitimate view. We will talk more about this in chapter four.

Today it is catastrophism within the framework of uniformitarianism – but not the global catastrophe of Genesis – which is the predominant view in modern geology. It is now acknowledged that catastrophes have played a huge part in our earth's past – but not the global flood/catastrophe of Genesis. Christians have held for centuries that a world-wide catastrophe (the Genesis Flood) easily explained what was observed in the rocks and landforms, but because God was involved in that view, it was not allowed by the scientific community as a legitimate explanation for the geological features all around us. In the early 1970s as I studied geology, if I had used the word catastrophe in one of my college papers, I would have been flunked! That word catastrophe smacked of the Bible.

As stated earlier, what one thinks of Genesis is too important to compromise. When it is compromised, God disappears, as well as the need for, and a plan for, man's salvation. Uniformitarianism denies God's involvement in His creation. The record of Genesis is very clear that God was involved. A Christian, if he/she is going to remain consistent in his/her faith, simply cannot have it both ways. In the development of a complete and consistent worldview, the pieces of the puzzle of life have to fit together. The uniformitarian view of earth history does not allow us to accomplish this. By definition it is a worldview that leaves God, at least the Biblical God, out of the picture. It ascribes the geological history of the earth to totally naturalistic and unguided processes. The Bible is very clear that God not only created the world and everything in it, but also watches over it. The Psalms are replete with this theme. The first chapter of Hebrews echoes this by saying, *"He upholds all things by the word of His power."* He is involved! When it comes down to it, uniformitarianism is either a Deistic belief about God and is therefore not consistent with an orthodox view of God. It ascribes false things to the God of the Bible. Or it is an atheistic view: God does not exist, and so was not involved in creation.

The main point to take away from the geological revolution in the 1800s is that a philosophical shift in worldviews is what drove the development of modern geology and the move away from a straightforward reading of Genesis!

Chapter Three
Uniformitarianism vs. The Biblical Account

What is Uniformitarianism?

Uniformitarianism is the root of modern scientific thought. It is the idea that the past geology of our globe (and the universe by extension) can be explained by present observable geological processes operating over long periods of time without recourse to a Biblical Flood or intervening God.

In 1785, before examining the evidence, the Deist James Hutton, the Father of Modern Geology, proclaimed:

> *The past history of our globe must be explained by what can be seen to be happening now...No powers are to be employed that are not natural to the globe, no action to be admitted except those of which we know the principle.*[2]

The doctrine of uniformitarianism is not a scientific refutation of the Bible. It is an alternate, unprovable idea of earth history in conflict with the Bible. It is a totally naturalistic explanation for the history of our earth and the universe. This is in direct contradiction to what the Bible clearly teaches throughout its pages – that God created the earth and universe in six 24-hour days about 6,000 years ago, flooded it, because of sin, and continues to monitor the universe and intervenes where He deems best. In the above-mentioned book by Hutton, he said he viewed the history of the earth as cyclical – no apparent beginning and no apparent ending. This view was not uncommon in his day, although it was a sort of "underground" belief, not held by the prevailing scientific community. Remember, the majority of the scientists of Hutton's day accused him of promoting atheism.

Uniformitarianism was the logical outcome of a Deistic belief about God – that there may have been a Creator, probably there was, but if there

---

[2] Hutton, J., 'Theory of the Earth,' a paper (with the same title of his 1795 book) communicated to the Royal Society of Edinburgh, and published in Transactions of the Royal Society of Edinburgh, 1785; cited with approval in Holmes, A., Principles of Physical Geology, 2nd edition, Thomas Nelson and Sons Ltd., Great Britain, pp. 43–44, 1965.

was, He left the creation to govern itself and to run on its own. Therefore, He did not and does not intervene in the physical universe with things like floods and other disasters. However, because uniformitarianism was framed within a scientific framework, it became accepted as legitimate and the only scientific view of earth history today. It was the expected child of The Enlightenment (1600s-1800s) – a philosophical shift in man's view toward God, himself and the world around him. In the historical development of uniformitarianism, it is important to realize that the philosophical shift came first, not the geological evidence. In science, it became the interpretation of the geological evidence. The Geologic Column, a sort of time scale for earth history, was developed as an evolutionary, uniformitarian interpretation of the observable rock record.

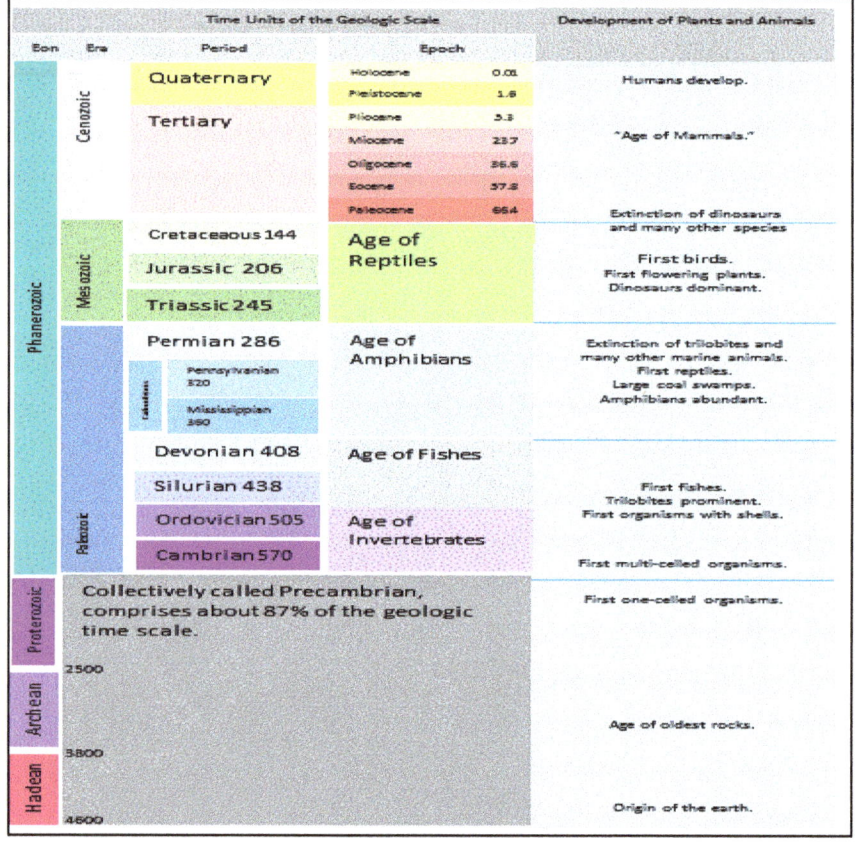

This is the basic geologic column showing the year in which the geologic periods were formulated. The ages for these various periods were added soon after, long before radiometric dating was even conceived.
Ages are in millions of years.

Uniformitarianism has now dominated geology for the past 200 years. As the doctrine of uniformitarianism took hold in geology, the conclusion for the rest of science was predictable. Biology soon followed with a uniformitarian, long ages view of the development of biological life – including man. Most of science, especially the historical sciences (those scientific disciplines that seek to explain origins and earth history in terms of uniformitarian ideas) have been influenced since then. Some of these include:

• **Anthropology** – the study of ancient man. Ancient man is viewed as primitive and therefore, less developed than modern man. Modern man is the culmination of millions of years of gradual evolution, and is therefore superior to ancient man. Religion and a belief in a God are viewed as a product of that development. The quest to find the "missing link" in man's evolution dominates the field of anthropology today.

• **Archaeology** – the study of man's development through the study of ancient artifacts. Originally a high value was placed on the Bible and ancient artifacts were interpreted in light of that history. Increasingly, however, secular histories and chronologies from other cultures replaced the Bible as interpretive guides. All discoveries since the early 1900s have been interpreted accordingly. One of the outcomes of this was embracing a particular view of Egyptian chronology which would have begun many years before the Genesis Flood. All discoveries are interpreted in light of these secular conclusions. The Bible is regularly criticized and discredited. The modern magazine, Biblical Archaeology, constantly attempts to "correct" or even contradict the Bible's view of man's history. I have subscribed to this magazine in the past and can tell you that this is the trend, not the exception.

• **Paleontology** – the study of fossils. In light of evolutionary biology, a uniformitarian interpretation of fossils became the "proof" for evolution. As the Geologic Column, a uniformitarian interpretive time scale, was developed and modified beginning in and throughout the 1800s, all fossil discoveries since have been interpreted in light of that Geologic Column.

• **Psychology** – the study of the soul. If man evolved, then his soul evolved too. All of man's problems were now viewed in terms of his naturalistic, evolutionary development. The causes of his emotional problems were viewed as environmental and physical. Morality became a relative term with no absolutes and of no real consequence. The Fall of man, a Biblical idea, was no longer seen as a historical event with eternal consequences, but as harmful religious dogma.

• **Astronomy** – the discovery and recording of stars and planets in the universe. The naturalistic explanation for the origin of the universe continues to be one of the main goals of modern astronomy. Because the Big Bang fits the best to a naturalistic explanation of the universe, it is the dominant "scientific" view of today. Another product of uniformitarian astronomy is the search for life on other planets. Because man is viewed as a naturalistically evolved creature, then evolution must be going on in other worlds too. This is contradictory to the Biblical view that man was created special, in the image of God, unique from other creatures and was placed on a planet specifically designed for man's habitation.

• **Radiometric dating** – the practice of using radiometric decay as a way of determining ages for ancient artifacts, whether they be rocks or pottery. Radioactive decay has become a sort of proof for an old earth and universe and any ages derived from this method that are not in line with a uniformitarian view of earth history are thrown out as invalid. It is not readily admitted today that the reliability of radiometric dates is subject to the Geologic Column and a uniformitarian view of earth history. Any "young" dates derived from this method (and there are many!) are discarded. (For an example of this, see geologist Dr. Steven Austin's research on the recently-formed dacite lava from Mt. St. Helens, available from www.icr.org).

The new dome within the crater of Mt. St. Helens

Today modern science is a well-mixed concoction of real science and false science. However, the average person is rarely able or equipped to tell the difference. The scientific community has taken on the role of a sort of high priest who interprets the natural world around us and makes pronouncements accordingly. We ordinary humans are not allowed to question these conclusions. When establishing a geological "fact" as legitimate, it is necessary to cite scientists' research, studies and opinions – who embrace a uniformitarian view. This somehow gives the uniformitarian idea the punch it needs to convince the average person. The church, by and large, has bought into this process and readily accepts the conclusions of modern geology. Genesis is reinterpreted in light of "scientific findings." This started in the late 1700s and continues to this day.

The Biblical Account
In light of the development of uniformitarianism as the dominant view in science today, and in light of its anti-Scriptural directions, it is important for us who believe the Scriptures to reinterpret the rocks and landforms around us from a different view. In contrast to the uniformitarian, naturalistic Geologic Column as a representative of time and chance, the first eleven chapters of Genesis present an events table or column consisting of a short period of time. Look at the following illustration.

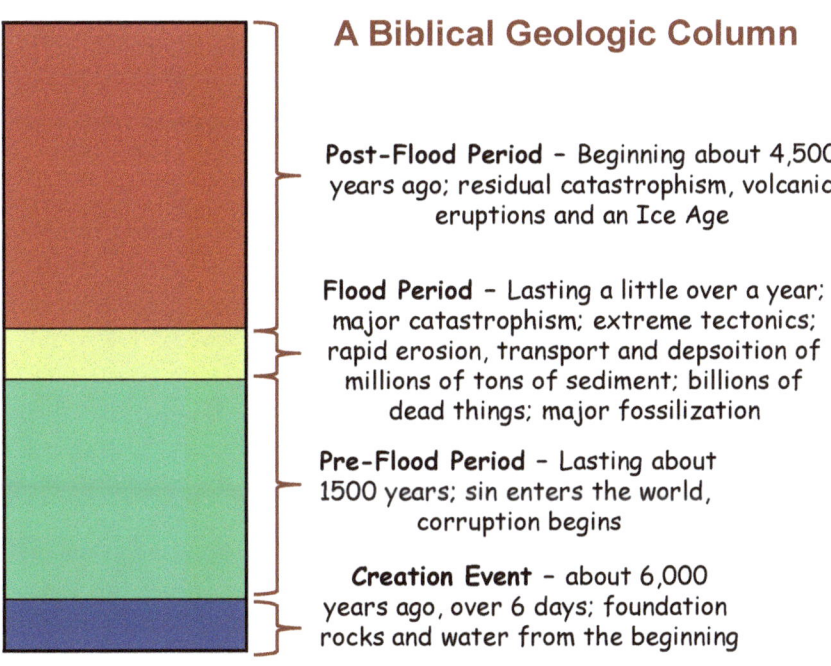

A Biblical Geologic Column

Post-Flood Period – Beginning about 4,500 years ago; residual catastrophism, volcanic eruptions and an Ice Age

Flood Period – Lasting a little over a year; major catastrophism; extreme tectonics; rapid erosion, transport and depsoition of millions of tons of sediment; billions of dead things; major fossilization

Pre-Flood Period – Lasting about 1500 years; sin enters the world, corruption begins

Creation Event – about 6,000 years ago, over 6 days; foundation rocks and water from the beginning

In this table, notice that the various sedimentary layers and geological activities of the Flood and Post-Flood Periods could be categorized and sorted within this hypothetical geologic column based on the Scriptural descriptions of the first eleven chapters of Genesis. This work is currently in progress among many Scriptural geologists today. See the bibliography for a few books and monographs on this subject.

Although much could be said for the Creation Event Period outlined in the illustration on the previous page, for purposes of this booklet, at this time, I want to concentrate on the following periods of geological events.

The Pre-Flood Period
Peter gives us a good, general picture of the world before the Flood.

> *2 Peter 3:3-7, 3 Know this first of all, that in the last days mockers will come with their mocking, following after their own lusts, 4 and saying, 'Where is the promise of His coming? For ever since the fathers fell asleep, all continues just as it was from the beginning of creation.' 5 For when they maintain this, it escapes their notice that by the word of God the heavens existed long ago and the earth was formed out of water and by water, 6 through which the world at that time was destroyed, being flooded with water. 7 But the present heavens and earth by His word are being reserved for fire, kept for the Day of Judgment and destruction of ungodly men.*

We can draw several significant geological conclusions from this passage of Scripture:

• In modern times the view of naturalism would prevail among mankind. Naturalism was born out of the religious view of Deism – God created but is not involved in His creation. Naturalism is an expansion of that view where God is totally irrelevant to the subject of earth history.

• The heavens came into existence by the word of God, not through naturalistic processes. The drift away from a God who created and has been involved in His creation leads ultimately to naturalism. The Scriptures teach otherwise.

• The earth was formed out of water and by water – not from a molten ball. One of the modern ideas of the beginning of earth history (which I learned as a kid) includes the development of earth from a hot molten blob which over billions of years cooled to form our planet.

- The earth at that time (the earth before Genesis chapter seven) was destroyed by water – a global flood of water. It is significant that Charles Lyell had as his personal goal to finally and totally eliminate the influence of Genesis and a global flood from the field of geology.

- There was a beginning to the universe, as described in Genesis chapter one, and there will be an end to the universe, brought about by the word of God, through fire.

Earth history and the fossil record should be viewed through these glasses. Therefore, the fossil record of life is one predominantly of life before the Flood. The mountain ranges of today were not the mountain ranges of the Pre-Flood Period. The volcanism of today was not the volcanism of the Pre-Flood Period. Today's oceans are not the oceans of the Pre-Flood Period. Today's environment is not the same as that of the Pre- Flood Period. In short, the world of Adam to Noah was completely different than it is today. The surface of that earth was totally erased by the great catastrophic flood of Genesis chapters 7-8.

The Flood Period
Genesis Chapters 6-9 are devoted to the Flood Period. For the sake of space, I will not print those chapters here. Please stop your reading at this point and refresh your memory by reading these chapters of Scripture. Several geological implications can easily be deduced from these chapters:

- Given the dimensions and construction of the ark, as recorded in Genesis, the vessel was obviously prepared for travel on water. Several books and DVDs have been released exploring this premise. You can select from a variety of these by going to www.answersingenesis.org, www.creation.com, or www.icr.org.

- In Genesis 7:11 we read,

> *In the six hundredth year of Noah's life, in the second month, on the seventeenth day of the month, on the same day all the fountains of the great deep burst open, and the floodgates of the sky were opened.*

The geological implications of this statement are enormous:

(1) Whatever "the great deep" means, "burst open" implies a tremendous amount of violent earth upheaval.
(2) Metamorphic rocks are thought to be formed from pressure and heat. The sort of friction, pressure and heat that would have resulted from this amount of earth movement would produce a great deal of metamorphic rock.
(3) The amount of sediment produced and marine life buried as a result would certainly, adequately explain the tremendous amount of sedimentary rock all over the world. It is a recipe for rapid and complete burial of billions of marine, plant and animal life.
(4) Given the break-up of at least the crust of the earth, the amount of volcanic activity and resultant lava produced would have been tremendous.
(5) Although it is not exactly known what "floodgates of the sky" means, it is obvious that this was a unique and significant geological, meteorological and environmental event. What we observe in the fossil record can be adequately interpreted by these events.

- In Genesis 7:19-22, we read,

> *19 And the water prevailed more and more upon the earth, so that all the high mountains everywhere under the heavens were covered. 20 The water prevailed fifteen cubits higher, and the mountains were covered. 21 And all flesh that moved on the earth perished, birds and cattle and beasts and every swarming thing that swarms upon the earth, and all mankind; 22 of all that was on the dry land, all in whose nostrils was the breath of the spirit of life, died.*

Whatever mountains existed prior to the Flood, the water covered them all. This might explain why we find billions of marine fossils at the tops of many mountains around the world. The other significant thing to notice is that there was a lot more water on the earth than existed before the Flood. Where did it go? I believe Psalm 104:5-9 gives an interesting answer to this question.

> *5 He established the earth upon its foundations... 6 Thou didst cover it with the deep as with a garment; the waters were standing above the mountains. 7 At Thy rebuke they fled; at the sound of Thy thunder they*

hurried away. 8 The mountains rose; the valleys sank down to the place which Thou didst establish for them. 9 Thou didst set a boundary that they may not pass over; that they may not return to cover the earth.

This clearly tells us that God was involved in forming *basins*, if you will; special places for the waters to drain into. Notice too that more significant geological activity (mountain building) took place as the "…mountains rose; the valleys sank down."

- As to the hardening of the layers upon layers of sediment produced by the Flood, notice Genesis 8:1:

    But God remembered Noah and all the beasts and all the cattle that were with him in the ark; and God caused a wind to pass over the earth, and the water subsided.

The total time for the Flood Period was about one year. It is evident from the significant sedimentary rocks around the world that these were laid down rapidly and in quick succession. This would definitely explain the flat contacts between layers of sedimentary rock so readily observed in the Grand Canyon.

In the following picture, one of many from the Colorado Plateau, you can see four separate deposits. Geologists call these geological events that record what has been deposited over millions of years. Each deposit is supposed to represent an environment or period of time that once existed. But between each deposit are two things: (1) flat contacts, indicating no erosion, and, (2) missing time; millions of years of missing time. These deposits could be interpreted as having been laid down rapidly and in quick succession in the first 150 days of the Genesis Flood. All that remains now is the remnant of what was scoured away during the last 171 days of the Genesis Flood.

According to a uniformitarian perspective, there are many layers of geological time missing with significant gaps between layers. Geologists call these **unconformities**. And they are a real geological mystery. Where did the missing layers go? Either they eroded over millions of years and consequently should show erosion events, or they are missing because the layers were actually laid down that way – rapidly and in quick succession. The phrase, "God caused a wind to pass over the earth, and the water subsided," catches my attention. Could the wind have contributed to the rapid hardening of the layers of sediment? It's an interesting idea. And the laying down and hardening of vast sedimentary layers of rock is still somewhat of a geological mystery. Was a unique chemical process involved in the hardening of these layers? At the very least, we see God intimately involved in His creation. This is certainly not a Deistic uniformitarian concept!

The next picture is of a formation located throughout the northeastern part of Yellowstone Park and the Beartooth Mountains which is called, in geology circles, The Great Unconformity. The top most part of the mountain consists of volcanic rock supposed to be 40 million years old. It rests directly on top of the band of rock called limestone, a sedimentary marine rock in the middle of the picture, supposed to be around 300 million years old. This band of rock in turn rests directly upon basement rock, consisting of granite and metamorphic rock supposed to be 2-3 billion years old. Obviously, there is a lot of missing time between these various deposits! Where did it go? The contact points of the three rock formations are flat, again, indicating no erosion. Would it not make more sense to simply conclude that these were various deposition events within the Genesis Flood that occurred rapidly and in quick succession of one another?

• In Genesis 8:2-3, we notice,

> 2 *"Also the fountains of the deep and the floodgates of the sky were closed, and the rain from the sky was restrained; 3 and the water receded steadily from the earth, and at the end of one hundred and fifty days the water decreased."*

All around the world the ocean floor exhibits rifts or cracks creating what geologists have called plates. From a uniformitarian perspective, these plates are interpreted as floating continents that were separated millions of years ago, from one main piece of land. This supposed uniformitarian geological event has become known as Pangaea. A reinterpretation would argue that the cracks are there because the fountains of the great deep burst open and closed again. How far the continents might have drifted, if they have drifted at all, is a uniformitarian idea, not necessarily a Biblical one. No one has scientifically demonstrated that the continents of long history of having moved from one part of the globe to the other. Taking the Genesis Flood into account, one could simply infer that the continents are the remnants of the carving up of the original pre-Flood land mass. One other significant geological event to notice here is the receding of the waters that had covered the earth. With the receding waters, we move into the Post Flood Period.

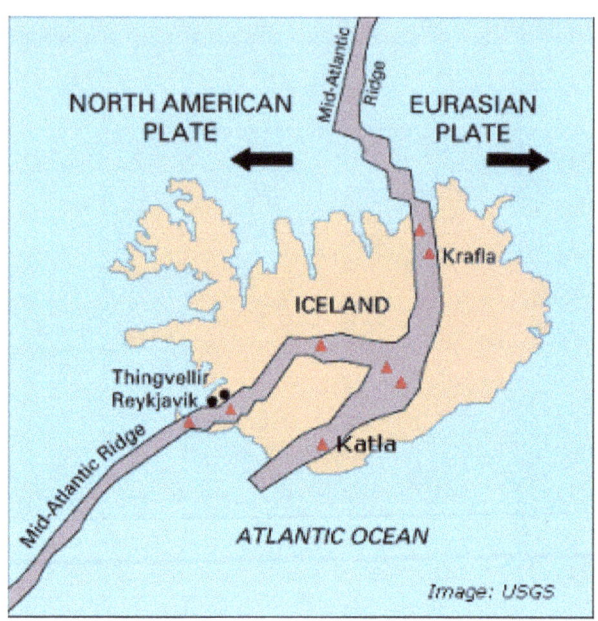

The Great Atlantic Rift

The Great Atlantic Rift runs right through Iceland and would be a remnant of the breaking up of the fountains of the great deep.

The Post Flood Period – Receding Waters; Volcanism and the Ice Age

During this much-overlooked period geologically, significant geological activity would have taken place. The millions of cubic miles of water rushing across the earth would have left tremendous scars and canyons. Planation surfaces – huge flat-topped mountains and plateaus – could have been formed during this period. Geologists readily acknowledge their existence, but their origin remains a mystery in the light of uniformitarian thinking. Events such as The Great Missoula Flood, discussed in chapter four of this booklet, an ice age, the creation of Lake Bonneville and Devils Tower could all adequately be explained by receding of the Flood waters and other Post Flood geological events. During this period significant volcanic activity, because of the bursting open of the fountains of the great deep, would have continued at least for a while, producing ash coverage which would result in cooler temperatures. Certainly, a Post Flood ice age (or ice event, as I like to say) is a reasonable outcome of these global conditions. I will discuss this more in chapter five – The Geology of Genesis. Virtually all of these events are interpreted within a uniformitarian framework in modern geology. We must put on a different set of glasses and reinterpret the geological record from a Biblical perspective. The explanations will be significantly different and will make sense!

Volcanoes like Mt. Shasta are a product of the much bigger event, the Genesis Flood. The amount of ash that was ejected from the many Flood and Post-Flood simultaneous volcanic eruptions is mind-boggling.

Chapter Four
Blinded for 200 Years –
The Story of J Harlen Bretz

J Harlan Bretz (1882-1981; he insisted that the J in his name be printed without a period!) was a geologist who started off teaching high school biology in Seattle during the early 1900s. During this time, he began studying the glacial geology of the Puget Sound area. As a result, he pursued and earned a PhD in geology at the University of Chicago in 1913. Subsequently he became an assistant professor of geology, first at the University of Washington and then at the University of Chicago. For the next several years he studied the fascinating volcanic features of the Columbia Plateau in Central and Eastern Washington. His story is one of the most significant in the field of modern geology. It showed (1) that research should precede scientific statements and, (2) just how blinded modern geology can be.

J Harlan Bretz

With so much scientific research accumulated over the past 200 years, it is hard to contemplate how modern geology could be wrong when it comes to the age of the earth. The present paradigm of the geological history of millions of years of gradual, naturalistic changes is one that is considered to be an established, unassailable fact in our culture. Surely hundreds of brilliant geologists all saying roughly the same thing have to be right.

Worldviews are powerful forces and there is no scientist that wants to be accused of letting his personal beliefs affect his judgment. The perception of the public about scientists is that they are truly unbiased. The image is portrayed that they constantly weigh their discoveries and conclusions against the observable data and known laws. But is this the reality of things? I am going to tell you a true story to illustrate just how scientists can be blinded by their preconceived ideas.

J Harlen Bretz was a geologist educated in the uniformitarian view of geology at the University of Chicago where he earned his Ph.D. in geology in 1913. He was fascinated with the geology of Washington. Beginning in the summer of 1922 and for the next seven years, Dr. Bretz conducted his field studies of the unusual erosion features of Eastern Washington State, now called the Channeled Scablands National Monument.

The Channeled Scablands
Bretz coined the term *channeled scablands* in 1923 to describe a tremendously large landscape spread across hundreds of square miles in Eastern Washington – cutting across the Columbia Plateau. The Columbia Plateau consists of a series of basalt lava flows (a dark fluid lava made up of iron, magnesium, and some other dark minerals) up to two miles thick and covering an area of about 64,000 square miles! These channeled scablands consist of straight-walled canyons; hundreds of water-carved channels called coulees. Coulees are deep gulches or ravines. A huge amount of water cut through the extremely hard basalt of the Columbia Plateau, and left "scars." This must have been a tremendous catastrophic event – a flood of "Genesis" proportions. Bretz proposed that at some time in the past over 500 cubic miles of water had to have been involved in the creation of this amazing geological feature. He estimated the width of this flood was about 100 miles. Even more startling was his calculation of a flood hundreds of feet deep with water moving at speeds up to 80 miles per hour! The largest of the coulees is Grand Coulee, 50 miles long and up to 1000 feet deep.

Basalt lava is black to gray and characteristic of volcanic flows such as those of Hawaii. It is a very hard rock! The Columbia basalt flowed for over 60,000 square miles and is a mystery for modern uniformitarian assumptions.

The Scablands, Eastern Washington: 60,000 square miles of basalt lava flows like the one pictured. These features caught the attention of Bretz. But what amazed him were the flat-bottomed coulees that indicated a huge amount of water had ripped through this area at some time in the past.

Eastern Washington is an arid, desert-like terrain. It is unlike Western Washington with its rainforest climate and lush green vegetation. The geological features of the eastern part of the state initially baffled Dr. Bretz. Where did the water come from? From all indications to Bretz, the features had been carved by vast amounts of water. But what was the source? Although he was unsure of the origin of the huge amount of flood waters needed to carve the canyons and coulees, he did propose initially that it had come from an ice sheet somewhere to the north. Later the research indicated that the water had actually come from a breach in an ice dam near present day Missoula, Montana. Bretz at first called this event, the Spokane Flood. It is now called the Missoula Flood. Those in the geological community at the time had another derisive name for it – the **Bretz Flood**. Another reason this idea puzzled Bretz was that although the area seemed to require a cataclysmic water flow to form the geological features and to have carved through the hard basalt flows, this went counter to his uniformitarian training which demanded a slow, gradual erosion of these areas.

Bretz's paper had been published in 1923, arguing that the channeled scablands in Eastern Washington were caused by massive flooding. The entire geology community vehemently rejected his ideas, probably for the following reasons: (1) Bretz was a virtual unknown to the geological community. Who was he to be making such outrageous claims? (2) Until Bretz had conducted and reported his research, this area of Washington had not been thoroughly studied. The landscape was virtually unknown to the geological community. (3) Bretz's ideas were totally contrary to what the geological community was prepared to accept. Bretz had stated that the channeled scablands argued for a catastrophic explanation, not an uniformitarian one. This was preposterous! In short J Harlen Bretz was branded a heretic of sorts. Debate over the origin of the scablands went on for over 40 years!

In 1927 Bretz was invited to speak before the Geological Society in Washington D.C., where he was verbally ambushed. Later that year, Bretz commented in a subsequent rebuttal,

> *I think I am as eager as anyone to find an explanation for the Channeled Scabland of the Columbia Plateau that will fit all the facts and will satisfy geologists. I have put forth the flood hypothesis only after much hesitation and only when accumulating data seemed to offer no alternative.*

There were several geologists who countered with papers of their own – despite the evidence. The geology community continued to be resistant for many years until beginning in the 1950s. The advent of a new technology involving aerial photography was beginning to prove Bretz's ideas correct. Finally, in the 1970s his work was given full acceptance and Bretz's flood catastrophe eventually became a national monument – The Channeled Scablands National Monument. As I have traveled through this wonder of wonders, I have been awestruck. It is quite a site as you stand in the middle of one of the channels carved in a matter of hours. It gives me chills just standing there and imaging a wall of water, hundreds of feet high rushing toward me at speeds of 65-80 miles per hour!

An Ice Dam

A word about the ice dam in Western Montana – if you remember from chapter three, I talked about the climatic changes which must have occurred as a result of the Genesis Flood. One of these changes involved the huge amounts of ash spewed into the atmosphere with so many volcanic eruptions. The combination of the ash, cooler atmosphere and abundant moisture evaporated from warmer oceans would have produced ideal conditions for a sort of ice age. The ice dam in Western Montana, therefore, would have been leftover as the volcanic eruptions subsided, the oceans cooled and the snow and ice began to melt. Again, the Genesis Flood offers a great explanation for this geological, catastrophic, flood event.

If geologists are as unbiased and open as our culture thinks they are, why did they not accept Bretz's work? I think the facts of this story rather speak for themselves! The geological community had been so entrenched in a uniformitarian way of thinking; wearing a set of uniformitarian glasses that they just could not see it. This story should cause us to stop and think. Are we allowing ourselves to be influenced by a uniformitarian way of thinking? We must remain faithful to a straightforward reading of Genesis. Geological ideas will come and go depending on the set of glasses worn. But the Scriptures have been spoken by the eternal God and therefore do not change with the advent of new ideas and the latest cultural shifts. We must make sure we are wearing the glasses of Genesis when encountering contrary ideas to the words spoken by God.

Uniformitarianism, although not a hot topic in college Freshman Geology these days, nevertheless has shaped all of modern geology from the late 1700s on. When I studied geology in the 1960s and 1970s, it was the only view permitted in the study of geology. Despite its repackaging to include limited catastrophism, uniformitarianism is still a naturalistic philosophical view of the history of the universe and the framework for interpreting the rock record. It stands in direct contrast and contradiction to the Biblical view of our universe. We must be on our guard to recognize when uniformitarian influences are affecting our interpretation of the Bible and the world around us.

Chapter Five
The Geology of Genesis - Putting on the Right Glasses

Throughout the first part of this booklet I have talked about the important part that worldviews play in shaping our interpretation of the rocks and fossils around us. To review, a worldview is the way we see the world around us. It is what determines how we think and the decisions we make. Whether we want to admit it or not, our worldview also determines how we interpret the landforms and the rocks all around us. Everyone has a worldview. We may not be aware of it, but we all have one nevertheless. I liken my worldview to the philosophical glasses I wear. When I put on the glasses of uniformitarianism, I will see the rocks and fossils in a certain way. I can be convinced that there can be no other way to view them. I can organize the data around this outlook and make it logical and convincing, without even being aware that I even have a worldview. Remember, rocks, fossils and landforms do not come with dates or ages. These are determined by my "glasses." The glasses of uniformitarianism have dominated the field of geology for the past 200 years, telling us that the only legitimate view in reconstructing earth history is a purely naturalistic one. I am going to ask you to take those glasses off and put on the glasses of Genesis – specifically the glasses of the catastrophic global flood of Genesis. When you do this, you will see amazing things! When I did this in 1972, my whole perspective changed. For the first time in my life I was seeing the world in a totally new and different way – and one that made sense in light of what I was reading in Scripture.

The Purpose for a Global Flood
In Genesis 6:5-7, 17, we read,

> *5 Then the LORD saw that the wickedness of man was great on the earth, and that every intent of the thoughts of his heart was only evil continually. 6 And the LORD was sorry that He had made man on the earth, and He was grieved in His heart. 7 And the LORD said, "I will blot out man whom I have created from the face of the land, from man to animals to creeping things and to birds of the sky; for I am sorry that I have made them... 17 And behold, I, even I am bringing the flood of water upon the earth, to destroy all flesh in which is the breath of life, from under heaven; everything that is on the earth shall perish."*

A straightforward, unbiased reading of this passage makes it clear that God's intention was to destroy all human beings, etc., on the earth – not just in a certain small part of the world.

Evidence of a Global Flood

As one examines the fossil record, one of the first things that he/she will notice is the destruction of land dwelling life and plant life as well as marine life on a massive global scale. Whether one chooses to accept this or not, the fossil record agrees with the statement of these passages of Scripture. Those who live in the neighborhood of southwestern Wyoming can travel a short distance to the hundreds of square miles of layer after layer after layer of limestone (a sedimentary rock), consisting of billions of exquisite fossil fish, insects, bats and other varied animals, including crocodiles, palm trees, sharks and so on. A few years ago, my son and I spent a couple of days digging in this area. As we carefully separated the layers of sandstone, we were amazed to see literally hundreds of fossils, in some cases perfectly preserved, laid out as if instantly buried. Some of these fossil creatures were in the act of excreting when they were suddenly buried! My son found a large mosquito the size of a silver dollar, perfectly preserved! These creatures were obviously buried in a hurry and under conditions that do not exist today – at least on that scale! Southwestern Wyoming is not the exception either. There are sites like this, too numerous to list here, all around the world. These massive fossil deposits have been appropriately called ***graveyard fossils***.

The author standing in front of some of the amazing fossils that have been found in southwestern Wyoming

The Geological Implications of a Global Flood
Genesis 7:11-12 states,

> *11 In the six hundredth year of Noah's life, in the second month, on the seventeenth day of the month, on the same day all the fountains of the great deep burst open, and the floodgates of the sky were opened. 12 And the rain fell upon the earth for forty days and forty nights.*

The geological implications of just these two passages are enormous. Look at these two verses again. At the very least they convey an event of significant earth movement and great amounts of water being released – but not just through rain. There might have been great amounts of water stored in the earth and released during this upheaval. With great amounts of earth movement come heat, friction and lots of potential sediment. Much of the metamorphic rock we observe today could have been formed during this period and immediately after as the movement of earth subsided and settled out. The word metamorphic means **change in form**, and metamorphic rocks are thought to be formed through tremendous amounts of pressure and heat. Metamorphic rocks are characterized by twisting, swirling and layered sorting of the minerals that were at one time jumbled together in other kinds of rocks. Many metamorphic rocks look as if my mother's chocolate chip peanut butter brownies had been re-cooked and stretched into a new type of desert.

An example of a metamorphic rock: notice how the minerals that form the rock have been arranged in segregated bands of alternating light and dark colored minerals.

In addition to earth movement, the tremendous energy and heat from magma thought to be a part of the mantle of our earth would have broken through fissures and cracks to release great amounts of magma in the form of lava. The earth is literally dotted with thousands of extinct, dormant and active volcanoes. Stretched across Washington, Oregon and Idaho is one of the largest basalt (a dark lava rich in iron and magnesium) flows in the world. This is referred to in geology as the Columbia Flood Basalts. These kinds of geological activities are not taking place on this scale today. It is obvious that some kind of significant global catastrophe took place not so long ago in our earth's past. The observed data fit well with what Genesis states. I think it is interesting that geologists estimate that 75% of the earth's rock formations are sedimentary in nature. Sedimentary rocks are rocks formed from hardened layers of mud, clay and other sediments that at one time were laid down by water. It is also true that a vast area of the earth's surface, including the ocean floor, is covered with volcanic rocks. These two kinds of rocks, sedimentary and volcanic, would be precisely what one would expect, if indeed there had been a significant breaking up of the earth, as Genesis records.

Thousands of square miles of sedimentary shale, sandstone and limestone in the Grand Canyon speak of a huge watery catastrophe.

The huge sandstone formations in Monument Valley are a tribute to the power of the receding Genesis Flood. As far as the eye can see, there are sedimentary rocks that dwarf anything and anyone around them. This valley was, before the receding Flood waters, filled with sediments laid down during the first 150 days of the Flood. The monuments stand as a stark reminder of the worldwide nature of the Genesis Flood.

Unimaginable amounts of volcanic tuff and ash that form The Grand Canyon of the Yellowstone speak of tremendous volcanic eruptions, the likes that we have not seen in modern times!

Climatic Changes Produced by a Global Flood – An Ice Age!

In addition to the significant changes in Earth's existing landforms and the formation of new ones, there would obviously have been drastic climatic changes. The fossil record of plant and animal life indicates a far different climate than that of today.

As more and more magma extruded and mixed with the abundance of water released on the earth during and after the height of the global flood, there undoubtedly would have been a temperature change in the waters. Relatively warmer water would have produced more evaporation and condensation. As volcanic activity increased at least for a while during and after the flood, the release of significant amounts of ash would have clouded the skies and produced a cooling in the atmosphere by inhibiting the sun's radiation. This in turn would have caused the abundance of moisture released into the atmosphere from the warmer oceans to fall back to Earth in the form of snow – lots and lots of snow! The conditions implied in Genesis would have been ideal for a sort of an ice age. Evidence shows us that the northern and southern hemispheres were once covered with ice sheets. Today only vestiges of these remain in the form of polar ice caps. Did it really happen this way? We don't really know, but this scenario is not out of line with the implications expected from these passages of Scripture. One of the areas of greatest puzzle for geologists today is the ice age (or ages, depending on whom you talk to). Scientists today are as baffled as they have ever been about what kinds of conditions would have produced the mechanism for an ice age. It is still one of the greatest geological mysteries of our day.

One of the most amazing features of the Beartooth Mountains in Montana are the glacial troughs or valleys that stand out as evidence of the once massive glaciers that had occupied this area in the past. This classically U-shaped glacial valley of Rock Creek Canyon is over a thousand feet deep. The depth of the ice would have exceeded this, as it also sculpted the top of the Beartooths.

In 1815 an awesome volcanic explosion erupted from Mt. Tambora in Indonesia. It had the greatest effect on the American northeast, New England, the Canadian Maritimes, Newfoundland, and northern Europe. The effects of the eruption continued for several years. Encyclopedia Britannica reports,

> *In 1816, parts of the world as far away as western Europe and eastern North America experienced sporadic periods of heavy snow and killing frost through June, July, and August. Such cold weather events led to crop failures and starvation in these regions, and the year 1816 was called the "year without a summer."*[3]

All this is to say that the Genesis Flood and its aftermath offer the best explanation for a mechanism that could have caused an ice age. What would have been the effect of hundreds of volcanic eruptions produced by the catastrophic flood of Genesis and its aftermath?

One of the greatest volcanic remnants in the United States is the Absaroka Range of mountains. This vast chain of eroded volcanoes stretches from the north side of Yellowstone National Park, around to the northeast side and down the east side of Yellowstone. Geologists estimate that there is over 9,500 cubic miles of volcanic lava, tuff and ash that make up these mountains! To give you an idea of just how much we are talking about, the eruption of Mt. St. Helens in 1980, as destructive as it was, produced only .25 cubic miles of volcanic debris by comparison. The amount of volcanic activity we see today pales when compared to the volcanic eruptions the occurred during and shortly after the Flood.

Pilot Peak and Index Peak, just northeast of Yellowstone National Park, part of the massive Absaroka volcanic chain

---

[3] Encyclopedia Britannica. Found at http://www.britannica.com/EBchecked/topic/581878/Mount-Tambora.

Notice the comprehensive global destruction and geological upheaval of Genesis 7:18-24:

> *18 And the water prevailed and increased greatly upon the earth; and the ark floated on the surface of the water. 19 And the water prevailed more and more upon the earth, so that all the high mountains everywhere under the heavens were covered. 20 The water prevailed fifteen cubits higher, and the mountains were covered. 21 And all flesh that moved on the earth perished, birds and cattle and beasts and every swarming thing that swarms upon the earth, and all mankind; 22 of all that was on the dry land, all in whose nostrils was the breath of the spirit of life, died. 23 Thus He blotted out every living thing that was upon the face of the land, from man to animals to creeping things and to birds of the sky, and they were blotted out from the earth; and only Noah was left, together with those that were with him in the ark. 24 And the water prevailed upon the earth one hundred and fifty days.*

This historical record describes in detail the global catastrophe known as **Noah's flood**. Note the words, "...all the high mountains everywhere under the heavens...all flesh...perished...every swarming thing...and all mankind...all that was on dry land...all...blotted out every living thing...only Noah was left...." I think one could safely conclude that this catastrophe was global! Also, the amount of mud and clay sediments that would have been stirred up and buried with layer upon layer of plants and animals would have been enormous – millions of tons of sediments worldwide. And the abundance of sedimentary formations around the world attests to this.

The Geology of Receding Global Flood Waters
Chapter eight of Genesis begins to describe the subsiding or receding stage of the flood. We touched on this briefly in chapter three, but let's look at this section again (Genesis 8:1-3).

> *1 But God remembered Noah and all the beasts and all the cattle that were with him in the ark; and God caused a wind to pass over the earth, and the water subsided. 2 Also the fountains of the deep and the floodgates of the sky were closed, and the rain from the sky was restrained; 3 and the water receded steadily from the earth, and at the end of one hundred and fifty days the water decreased.*

From this passage the picture is created as one of tremendous amounts of water rushing off of the earth and draining – but where to? Where did the water go? According to Psalm 104:6-9, the water receded into basins prepared by God;

> *6 You covered it with the deep as with a garment; the waters were standing above the mountains. 7 At Your rebuke they fled; at the sound of Your thunder they hurried away. 8 The mountains rose; the valleys sank down to the place which You established for them. 9 You set a boundary that they may not pass over; that they may not return to cover the earth.*

Where did all the water go from the Flood? Apparently, as the valleys sank down, the ocean basins were formed by God through His word. Again, this is not a Deistic uniformitarian idea. The Scriptures teach that God was involved in an active way in the geology of the earth during Noah's day.

Rapid Mountain Building
To illustrate how mountains can rise rather quickly, here is an interesting bit of volcanic history. On May 8, 1902, the incredible volcanic eruption of Mt. Pelee on Martinique took place. One of the curious side occurrences of this eruption was the creation of a huge mountain. In October of 1902 a lava dome began to rise out of the crater floor. Over the course of a year it grew over 1000 feet above the base of the crater floor. It has been described as the most spectacular lava dome produced in historic times. It was 350 to 500 feet thick at its base and rose remarkably at 50 feet per day! This became known as the ***Tower of Pelee***. At its maximum size, it was twice the height of the Washington Monument and equal in volume to the Great Pyramid. It finally collapsed in 1903 after 11 months of growth. A very similar thing took place at Mt. St. Helens in Washington. The geology of our earth does not require millions of years to form and shape landforms.

Even secular geologists agree that most of our mountain ranges of today were formed around the same time in the last 65 million years of geologic history. Although I do not agree with the time frame here, it is worth pointing out that 65 million years is roughly 1% of Earth's geologic history according to secular geologic reckoning – so it is relatively recent and rapid in the secular scheme of things.

Planation Surfaces – Mysteries in Uniformitarian Geology

All over the earth are landforms called in geology, planation and plateau surfaces. Planation surfaces are large flat areas of the earth, not accounted for by slow gradual erosion. Some of these surfaces are many miles across. Many of these are located on the tops of very high mountains. I have observed these at the tops of some of the Beartooth Mountains near Cooke City, Montana. Planation surfaces are well known among geologists but their formation remains a mystery in modern geology. But if large amounts of water receded at great speeds across a landscape, it is easy to understand how huge areas of land could have been sheared off and their contents deposited elsewhere. An example of this is in western Montana. On the tops of many of the planation surfaces in this area of Montana are scattered rounded quartzite boulders and cobbles of varying sizes. Quartzite is a very hard metamorphic rock made up of – you guessed it – quartz, a hard mineral. Many of these boulders have percussion marks on them created as the boulders were banging against each other, moved along by surging, receding waves of huge amounts of water. The amount of force needed to leave indentations in these hard rocks and rounding them would have been tremendous. The other interesting thing here is that these quartzite boulders are not from Montana, but were moved from many miles east and northeast of this area and then deposited in Western Montana.

The Grand Canyon is an amazing place. But most people miss the real significant geologic landform of the planation surfaces, pictured here at the top of the picture. The Grand Canyon is really a plateau with a huge gash in it. But what geologic force would have eroded the canyon and yet leave the top flat? The receding stage of the Flood with its shearing and channeling power easily explains this mystery.

The Power of Surging Flood Waters

Many years ago, I had the opportunity to visit Crescent City, California while on vacation with my family. This would have been shortly after the great Alaskan earthquake of 1964. This devastating earthquake created what were then called ***tidal waves*** (now called tsunamis). Crescent City was literally wiped off the map by a tidal wave created from this earthquake thousands of miles away. All over this city were concrete foundations with nothing on them and huge bridge pylons that had been moved off their bases by the force of the surging waves. Crescent City is now renamed, the ***Comeback Town***, as people have returned and rebuilt the city. Crescent City is just one example of what the powerful force of water can do, and the global flood of Genesis has left its mark all over the earth.

The Right Glasses

All one has to do to see evidence of a global flood, is to put on the right glasses. There are so many more examples I could site, but space will just not permit it here. In the back of this booklet is a bibliography from which you can read about many, many more examples of the geology of Genesis.

Today Bible-believing Christians are viewed as foolish and ignorant. Contrary voices confidently assert that evolution is a fact demonstrated by the discoveries in geology, biology, paleontology, astronomy and radiometric dating. But in reality, the physical evidence favors the description found in Genesis. It took some re-education for me to view the rocks differently from what had been my outlook. But soon, looking at the earth from a Biblical perspective became second nature for me. Remember, no matter how strong the evidence may or may not appear to be, a person's world view will ultimately determine how he or she interprets the evidence. What glasses are you wearing? Have you taken stock of your worldview lately? Now may be a good time to do just that.

For Group Discussion

Chapter 1
1. Who were the two main *movers and shakers* in the formation and establishment of modern geology?
2. What were their contributions?
3. Discuss how these men's views influenced the view of Scripture of many Christians.

Chapter 2
1. What was The Enlightenment?
2. What were the main effects on scientific thinking?
3. How has The Enlightenment affected our modern thinking?

Chapter 3
1. What is uniformitarianism?
2. In what ways is this idea contrary to Genesis?
3. What are the main stages of a Scriptural geological events column?
4. Within each stage, what are the main geological events?

Chapter 4
1. How does the story of J Harlen Bretz demonstrate the influence that our worldviews have on our scientific reasoning about Earth history?
2. Many people go through life never defining their worldview. If you have never done this, use the space at the end of the book to outline your worldview. If you have a worldview, have you noticed any inconsistencies with the Scriptures? If so, what are they?
3. Many people experience a sort of conversion from a belief in evolution or uniformitarian geology to a trust in the Scriptures. If you experienced this, what were the main events in your journey?

Bibliography

The Genesis Flood, John C. Whitcomb, Jr., Henry M. Morris, 1961, The Presbyterian and Reformed Publishing Company, Philadelphia, PA

Scientific Studies in Special Creation, Walter E. Lammerts, editor, 1971, Baker Book House, Grand Rapids, Michigan

Coming to Grips with Genesis, Terry Mortenson, Thane H. Ury, editors, 2008, Master Books, P.O. Box 726, Green Forest, AR 72638

The Mythology of Modern Dating Methods, John Woodmorappe, 1999, Institute for Creation Research, El Cajon, CA

Newton's Revised History of Ancient Kingdoms, Larry and Marion Pierce, editors, 2009, Master Books, P.O. Box 726, Green Forest, AR 72638

In the Minds of Men, Ian T. Taylor, 1987, TFE Publishing, P.O. Box 5015, Stn. F, Toronto, Canada

The Faces of Origins, David Herbert, 2004, D&I Herbert Publishing, London, Ontario, Canada

Science According to Moses, G. Thomas Sharp, Volumes 1-3, 1992, Creation Truth Publications, P.O. Box 1435, Noble, OK 73068

A World Without Heroes, George Roche, 1987, Hillsdale College Press, Hillsdale, Michigan 49242

Scientific Anomalies and Other Provocative Phenomena, compiled by William R. Corliss, 2003, The Sourcebook Project, P.O. Box 107, Glen Arm, MD 21057

The Origin of Species Revisited, Wendell R. Bird, Volumes 1&2, 1989, Philosophical Library, Inc., 31 West 21st Street, New York, N.Y. 10010

The Young Earth, John Morris, 2007, Master Books, Master Books, P.O. Box 726, Green Forest, AR 72638

Radioisotopes and the Age of the Earth, Larry Vardiman, Andrew Snelling, Eugene F. Chaffin, editors, 2005, Institute for Creation Research, P.O. Box 2667, El Cajon, CA 92021 and Creation Research Society, 6801 N. Highway 89, Chino Valley, Arizona 86323

Lord Kelvin and the Age of the Earth, J.D. Burchfield, 1975, Science History Publishers, New York, NY

In Six Days, John F. Ashton, editor, 2000, Master Books, Master Books, P.O. Box 726, Green Forest, AR 72638

The Great Turning Point, Terry Mortenson, 2004, Master Books, Master Books, P.O. Box 726, Green Forest, AR 72638

The Geologic Column, John K. Reed and Michael J. Oard, editors, 2006, Creation Research Society Books, 6801 N. Highway 89, Chino Valley, Arizona 86323

New American Standard Bible, 1977, The Lockman Foundation, La Habra, CA

Picture Credits

Introduction
Big Horn River: Photo. Public Domain. Found at http://www.nps.gov/bica/photosmultimedia/North-District-photos-Gallery.htm, 6. Big Horn Canyon: http://scilearn-hscs2012.wikispaces.com/8A+Term+2, http://creativecommons.org/licenses/by-sa/3.0/legalcode, CC by-SA 3.0, 7. Henry Morris: Fair Use. Found at http://www.icr.org/article/2717/, 8.The Genesis Flood: Found at http://thenewcreationism.wordpress.com/2011/02/03/celebrating-the-genesis-flood/, 8. Creation v. Evolution: http://smithlhhsb122.wikispaces.com/Carlee+Little-Wilkins, http://creativecommons.org/licenses/by-sa/3.0/legalcode, CC by-SA 3.0, 10. Mt. St. Helens: Photo by Ewen Roberts. Found at https://www.flickr.com/photos/donabelandewen/6137917810/, : http://creativecommons.org/licenses/by-sa/3.0/legalcode, CC by-SA 3.0, 11. The Cascade Mountains: Photo by Rose Braverman. Found at https://www.flickr.com/photos/rose_braverman/6924717915/, http://creativecommons.org/licenses/by-sa/3.0/legalcode, CC by-SA 3.0, 11.

Chapter 2
Thomas Chalmers: Public Domain. Found at http://en.wikipedia.org/wiki/File:Thomas_Chalmers_by_David_Octavius_Hill,_c1843-47.jpg, 15. William Buckland: Public Domain. Found at http://www.redorbit.com/education/reference_library/science_1/palaeontologists/1112974178/william-buckland/, 15. Charles Spurgeon: Public Domain. Found at http://iperceptive.com/authors/charles_spurgeon_quotes.html, 16. Charles Hodge: Public Domain. Found at http://en.wikipedia.org/wiki/Charles_Hodge, 16. B. B. Warfield: Public Domain. Found at https://www.logos.com/products/search?q=Warfield&sort=rel&pageSize=60&Author=1054%7cB.+B.+Warfield&redirecttoauthor=true, 17. Cyrus Ingerson Scofield: Public Domain. Found at http://creation.com/genesis-13-undermines-gap-theory, 17. Scofield Study Bible: Found at http://www.betterworldbooks.com/scofield-study-bible-iii-nkjv-centennial-id-0195279654.aspx, 17. Gleason Archer: Found at http://davidsonpress.com/endorsements.htm, 18. Encyclopedia of Bible Difficulties: Found at http://www.goodreads.com/book/show/458027.Encyclopedia_of_Bible_Difficulties, 18. Terry Mortenson: Fair use. Found at http://www.answersingenesis.org/articles/wow/about, 19. The Great Turning Point: Found at https://answersingenesis.org/store/product/great-turning-point/, 19. James Hutton: Public Domain. Found at http://creationrevolution.com/james-hutton-the-man-who-found-time/, 19. Charles Lyell: Public Domain. Found at http://commons.wikimedia.org/wiki/File:Charles_Lyell.jpg, 19. James Hutton: Painting by Sir Henry Raeburn. Public Domain. Found at http://en.wikipedia.org/wiki/James_Hutton, 21. The Rock Cycle: http://scienkist.wikispaces.com/Rockcycle, http://creativecommons.org/licenses/by-sa/3.0/legalcode, CC by-SA 3.0, 22. Charles Lyell: Photo. Public Domain. Found at http://en.wikipedia.org/wiki/Charles_Lyell, 22. Principles of Geology: Found at http://www.goodreads.com/book/show/6552554-principles-of-geology-volume-1, 22. Young Charles Darwin: By G. Richmond. Public Domain. Found at http://en.wikipedia.org/wiki/Charles_Darwin#mediaviewer/File:Charles_Darwin_by_G._Richmond.png, 23. The Beagle: Public Domain. Found at http://darwinbeagle.blogspot.com/, 23.

Chapter 3
Geologic Column: Image by Vicki S Nurre, 26. Mt. St. Helen's Dome: Photo courtesy of USGS, 28. Biblical Geologic Column: Image by Patrick J. Nurre, 29. Colorado Plateau: Photo by Patrick J. Nurre, 34. The Great Unconformity: Photo by Patrick J. Nurre, 34. The Great Atlantic Rift: Courtesy USGS. Found at http://filesfromtoni.blogspot.com/2010/04/katla-next-icelandic-volcano-to-blow.html, 35. The Great Atlantic Rift: Photo by Andrea Schaffer. Found at https://www.flickr.com/photos/aschaf/4826268076/, http://creativecommons.org/licenses/by-sa/3.0/legalcode, CC by-SA 3.0, 36 Mt. Shasta: Photo by Patrick J. Nurre, 37.

Chapter 4
J Harlan Bretz: Photo. Public Domain. Found at http://www.nps.gov/iceagefloods/d.htm, 38. Basalt lava, Hawaii: Photo by Patrick J. Nurre, 40. The Scablands: Photo by Patrick J. Nurre, 40.

Chapter 5
Giant fossils: Photo by Vicki S. Nurre, 44. Metamorphic Rock: Photo by Patrick J. Nurre, 45. Grand Canyon: Photo by Patrick J. Nurre, 46. Monument Valley: Photo by John Meyer, 47. Used by permission. Grand Canyon of the Yellowstone: Photo by Patrick J. Nurre, 47. Beartooth Mountains Valley: Photo by Patrick J. Nurre, 48. Pilot Peak and Index Peak: Photo by Patrick J. Nurre, 50. Grand Canyon: Photo by Patrick J. Nurre, 53.

Notes

Patrick Nurre has been a rock hound since childhood and has an extensive rock, mineral and fossil collection, having collected from all over the United States. In 2005, he started Northwest Treasures, which is devoted to designing geology kits for schools. He conducts numerous field trips each year in Washington State to such places as the Olympic Peninsula, Mt. Rainier, Mt. St. Helens, the Channeled Scablands, Mt. Baker and Whidbey Island. In addition, he also gives field trips to the volcano loop of Oregon and California, Mt. Hood volcanic area (Oregon), the eastern badlands of Montana and Yellowstone National Park. He is a popular speaker at homeschool conventions, schools, and churches. Patrick currently co-pastors a church in the Seattle, Washington area.

If you would like to contact Patrick about speaking or field trips: northwestexpedition@msn.com
For a list of speaking topics: NorthwestRockAndFossil.com

Other books by Patrick Nurre: these are all also available with sample rock, mineral, and fossil kits at NorthwestRockAndFossil.com.

Rocks and Minerals for Little Eyes (PreK-3)
Fossils and Dinosaurs for Little Eyes (PreK-3)
Volcanoes for Little Eyes (PreK-3)
Geology for Kids (3-6)
Rock Identification Made Easy (3-12)
Fossil Identification Made Easy (3-12)
Bedrock Geology (high school)
Rocks and Minerals: The Stuff of the Earth (high school)
Volcanoes, Volcanic Rocks and Earthquakes (high school)
Fossils, Dinosaurs and Cave Men (high school)
The Geology of Yellowstone – A Biblical Guide
Geology and the Hawaiian Islands

www.ingramcontent.com/pod-product-compliance
Lightning Source LLC
Chambersburg PA
CBHW040233020526
44113CB00052B/2717